The Hollow Earth

And the Special Earth Core Theory
(S.E.C.T)

Gravity Anomalies and Climate Change for the
New Aeon
2232-4464 A.D

Frater Akiba

Introduction 3
Chapter 1 The Kant-Laplace Model...6
Chapter 2 The Sub-Nebular Theory and Planetary Formations 9
Chapter 3 The Expansion of the Kant-Laplace Model 13
Chapter 4 Perceptive Relativity...17
Chapter 5 Quantum Geometrics 19
Chapter 6 Mass and Equivalency..22
Chapter 7 The Formation of Ring Systems and the Roche Limit........32
Chapter 8 A Solution to the Angular Momentum Problem 34
Chapter 9 Earth as a Spherical Conductor 38
Chapter 10 Metallic Entropy and Polar Reversals 40
Chapter 11 Phasal Development and Electromagnetic Flux..............43
Chapter 12 Wave Propagation and Seismic Activities 45
Chapter 13 Polar Flattening and the Earth Ellipsoid 55
Chapter 14 Secular Variation and Magnetic Anomalies 57
Chapter 15 Auroral Ovals 59
Chapter 16 Tectonic Plate Revolution..61
Chapter 17 Magnetic Anomalies 66
Chapter 18 The Causation of Atmospheric Temperature...................69
Chapter 19 Ice Ages and the Melting Ice Caps.................................73
Chapter 20 The GCM...75
Chapter 21 Ozone Holes...78
Conclusion..82

Introduction

"...In science conventional wisdom is difficult to overturn. After more than 20 years some implications of plate tectonics have yet to be fully appreciated by isotope geochemists....and by geologists and geophysicists who have followed their lead." (Richard Armstrong, The Persistent Myth of Continental Growth, Austr. J Earth Science. (1991) 38)

Standard models of Earth core dynamics and chemistry involve intricate interactions between a solid inner core and a liquefied outer core. This model demands many assumptions be made and produces numerous paradoxical inconsistencies. A number of these inconsistencies can be traced back to a few unnecessary and unfruitful assumptions. A more simplified hypothesis is presented here that is based on early planetary formation processes. It relies upon consistent phasal development and is unrestrained by assumptions of a solid Earth core and current gravitational constraints.

A solid Earth core engenders the necessity to fix density parameters within gravitational bounds which do not appear to coincide with known thermal and fertility variations nor the velocity profiles of compressional (P) and Sheer (S) wave oscillations. For this reason there is a present need for a more powerful theory, one not restrained by conventional assumptions and one that is able to smoothly transition from these early planetary formation processes to the present. The theory used to simplify our understanding of Earth core dynamics and at the same time provide a valid alternative is found in my

Special Earth Core Theory (S.E.C.T).

Geophysical studies of the earth core have developed steadily over the past 80 years. At present we possess a large amount of information derived primarily from seismic techniques which include the measure of wave oscillations during earthquakes that determine refraction and reflection indices, seismic tomography, measurements of seismic anisotropy and teleseismic converted waves, and nonseismic techniques such as GPS measurements of crustal deformations. Despite this vast accumulation of knowledge, however, many rheological anomalies remain which a solid earth core is unable to sufficiently address. These include but are not limited to: rigid plates with narrow boundaries that overly a chemically homogenous and shallow mantle with simple lateral and radial temperature structures, tectonics that do not account for broadly deformed zones of the crust, melting anomalies along ridges, volcanoes away from plate boundaries, geomagnetic inconsistencies, and oceanic and mountain gravity anomalies. The current Lehman hypothesis fails to answer for many of these observed inconsistencies.

In this book we will concentrate primarily upon velocity and density inconsistencies with depth and provide an alternative viewpoint of the Earths inner core in connection with the cores known electromagnetic properties. We will also focus on early planetary formation processes and develop a plausible yet highly unconventional approach which may provide a viable alternative of thought and at the same time provide an answer to many of these observed complexities created by the Lehman hypothesis.

Rather than provide a detailed analysis of each of the

observed anomalies we will present an overview of how S.E.C.T is able to answer, not only each of the aforementioned problems, but also other anomalous features such as secular variation, ozone holes, Ice Ages, and Climate Change in general.

I believe the time has arrived for a scientific revolution in our earth sciences, one that is able to potentially initiate a long overdue paradigm shift in our scientific thinking. It is my hope that the Hollow Earth and (S.E.C.T) will draw attention to some of these areas in need of revision while at the same time provide an alternative approach through which others may proceed.

Chapter One

The Kant-Laplace Model

The first semblance to a modern theory about the origins of our solar system is generally attributed to the German philosopher Immanuel Kant in 1775. Kant's theory contained a theme which described these early planetary formation phases as being the development of an early cloud-like distribution of chaotic matter that slowly coalesced in time. Kant's primary assumption was that mutual gravitational attraction caused particles and debris to begin moving and in these movements to subsequently collide. At this point chemical forces began bonding certain materials together. As some of these aggregates increased in size they eventually formed planets.

Some 40 years later the French mathematician and amateur astronomer, Pierre Simon Laplace, penned a treatise dealing with celestial mechanics. In his appendix he made a number of suggestions regarding the origin of our solar system.

Laplace began his theory by assuming that the sun had already been formed and that its orbital velocity and atmosphere were well established. As the sun then began to radiate away its energy it also began to contract thereby increasing its rotational velocity. Through electrostatic emulsion (centrifugal acceleration) the excreted matter eventually began planetary formation processes. The balancing act between centripetal and centrifugal forces then left a ring of debris behind in the equatorial plane of the sun. As the process continued each ring eventually produced a planetary body.

The Laplace model makes it necessary for planets to revolve around the sun in the same plane and in the same direction as the axial angular rotation of the sun.

Ultimately Kant's early accretion stages, and Laplace's planetary formation within the solar atmosphere, combined into what became known as the "Kant-Laplace nebular hypothesis". The basic structure of this hypothesis has stood for well over 100 years as the standard theory.

During the course of the past 100 years serious challenges to this theory have emerged and subsequent investigations have led many astronomers to discover contradictions within this structure. These include, but are not limited to, the discovery of asteroids with highly eccentric orbits, the retrograde orbits of planets such as Venus and Uranus, as well as some planetary satellites. Another problem discovered in this theory is the fact that the sun, while possessing over 99% of the solar systems mass, only carries 1% of its angular momentum. It was discovered that, for the solar system to conform to the Kant-Laplace model, either the sun should be rotating faster or the planets should be orbiting slower. Because of this apparent contradiction several scientists in the early 20th century decided that the Kant-Laplace model was no longer tenable.

Further independent investigations have led many astrophysicists to believe that planetary catastrophism occurred when our sun had a close encounter with another star. Significant developments along this line of thought have taken place since the 1960's utilizing a more detailed analysis of the processes involving stellar constitutions and their formation processes. From this more detailed analysis, and as a result of sophisticated observations, a new branch of helioseismology emerged which has dramatically shifted our scientific thinking regarding the solar and planetary formation processes leading to an abandonment of the solar catastrophe theory, however, newer models still tend to approach closer to the early Kant-Laplace theory.

Chapter Two

The Sub-Nebular Theory and Planetary Formations

The current approach used to explain the origins of our solar system deal with a gravitational collapse of stellar gas and dust. Investigators attempting to model the processes by which stars form begin with the assumption of a preexisting and highly protected cloud of dust with low interior temperatures that allow for matter accretion to be undisturbed. Any outside disruptions, such as ultraviolet radiation, must therefore be shielded from this clouds interior. This protection then allows for the central collapse of the condensation disc to occur rapidly while at the same time converting the interior matter into kinetic energy which in turn elevates the temperatures enough for thermonuclear fusion to begin.

While this theory has held ground for some time one major obstacle stands in its way: the Large Cloud of Magellan. Within the Magellanic cloud has been discovered nearly 50 luminous type O and B stars known as Shapley's constellation. Their age has been estimated to be no more than 20 million years with an average velocity not exceeding 10 km/s. It is quite apparent that, considering their age and velocity estimates, that none of these stars should have moved more than 200 parsecs from their current position. Considering this fact the Magellanic cloud should hold at least 60,000 solar masses of ionized hydrogen, however, instead one observes an entirely transparent region of space with no giant molecular complexes of hydrogen or cosmic dust. The obvious question we are forced to ask is, "How did these stars form when all the necessary prerequisites and gravitational constraints necessary for star building have been removed?"

The standard model of solar formation involves complex interactions between the gravitational force and the collapse of part of an interstellar cloud of gas and dust. Because of the orbital velocity of the cloud around the galactic center the more distant peripheral motions of the cloud are believed to be slower than the central parts, hence as the cloud collapses and begins its rotations, it must in some unknown manner conserve its angular momentum. This leads to centripetal contraction and a general flattening of the cloud. The final result is a disk of material around this central condensation.

The condensation disc is often referred to as a sub-nebula because it resembles the shape of a spiral galaxy.

Despite the fact that early formation processes of stellar bodies involve nucleation and fusion it is still largely believed that planetary formations within the solar atmosphere follow a similar process, thus the angular momentum from the sun is supposed to be captured by the planetary sub-nebula, and likewise, as the excreted matter of the solar mass is captured by the planetary sub-nebula, it is subsequently 'pulled' in toward the planetary central condensation disc where it is converted into a form of kinetic energy.

This marks the beginning stages of accretion supposedly

underlying the various thermal conditions and electrostatic equilibrium states necessary for the final physical outcome of a planetary body.

Thermal conditions are primarily assumed to be derived from the hot central mass of the sun which determines parameter distance control mechanisms believed to govern the relative orbital planes of the planets. Such a theory is used to explain why planets too close to the sun are unable to condense water from a gaseous state or why distances of planets such as Jupiter are able to form water ice. Largely it is believed that when a planet achieves a mass 10 times greater than the earth it is only able to retain larger amounts of the lighter elements such as hydrogen and helium contributing to low density indices.

The sub-nebula theory seeks to explain why outer accreting planets, such as Jupiter, are so large yet still possess such low densities. The subsequent acceptation of this theory, along with the variously interpreted thermal constraints, has led to the erroneous conclusion that only planets close to a star can be solid rocky bodies, while those at a distance of, say Jupiter, must be gaseous.

I believe it is necessary to reevaluate the three main constraints which define early planetary development. These include our current understanding of temperature control mechanisms for distance parameters, mass definitions, and density indices.

The acceptation of solar temperature dictates leads to the assumption that low density planets and larger mass planets must form at a distance of Jupiter and beyond, while smaller and more denser planets must remain confined to a close proximity to the solar mass. By reevaluating the current requirements for planetary formation processes, and reducing the necessity for invariant temperature constraints used to define planetary distances, in direct connection with strict mass and density definitions, we should be able to free up logic to take its natural course.

Despite the continuing justification of this rationale and the seeming simplicity of the sub-nebula theory it is unable to account for the neotetic observations of planetary bodies in orbit around other stars which do not conform to these predetermined constraints. In fact more recent discoveries have revealed larger planets than Jupiter in orbit around other stars at distances similar to Mercury.

Conventional wisdom dictates that scientific progress should improve with time. Theory should follow theory until there is a general consensus of thought well grounded in practical truth. Through such a process scientific rationale is able to advance in progressive stages by discarding old worn out hypothesis.

Our current understanding of planetary formation processes limits our scientific progress rather than providing a liberating freedom of forward and open dialogue. It isn't simply that the sub-nebula theory fails to account for observed formation anomalies rather it's continuing acceptance does not allow for any further forward momentum.

Chapter Three

The Expansion of the Kant-Laplace Model

The expansion of the Kant-Laplacian model deals primarily with the initial structure of a relativistic definition of planetary accretion phases by the inclusion of a quantum geometrical condensation. So rather than beginning from "mass" planetary formation processes begin from geometrical and volumetrically defined hyperstates of polarized space.

In this sense stellar and planetary formation processes originate prior to the utilization of mass as a necessary prerequisite towards fundamental creation. And since the outcropping of quantum geometrical matrices determines mass, space should not necessarily be defined as empty or void rather it should be conceived of rightly as it is: a tenuous substance of Unknown Origin which can be acted upon Hyperdimensionally through quantum geometrics.

The activity within these higher dimensions moves phenomenal space by creating polarized centers of vortextual spatial geometry while simultaneously causing the production of observable energy states. Each electrostatic equilibrium state will then subsequently determine the extent to which a planet or a star evolves.

Polarization of the medium is merely the observable outcropping we define as centrifugal and centripetal and this tends to lead to the erroneous supposition that somehow mass moves space when it is phenomenal space moving that produces the object we observe being moved. Such a reversal of relativistic logic excludes the necessity for pre-existing mass in the early development of planetary and stellar formation processes.

This process is no different than what one observes in the subnebulating formation found in atomic systems. At this level of phenomenon chemical affinity is not *gravitationally* confined and all subsequent attempts thus far to inculcate gravity into a Quantum state have failed miserably. For this reason, I believe it is necessary to consider a radical alternative, one that will be able to explain how solar and planetary formation processes can emerge from Quantum geometrical states excluding a necessity for mass and relativistic definitions.

Atomic structures do not necessitate a hidden force to provide substance to the already well understood attraction and repulsion one observes there. Electrons are bound to protons, protons to neutrons, and most of the mass density is concentrated at its nucleus. It follows then that the formation of molecular systems are determined by similar chemical affinities which compound into larger systems such as the complex molecular structure of human beings and cosmological bodies. Following out this simple reasoning and logic leads to the conclusion that larger systems merely accentuate the same internal chemical affinity sequences we observe at an atomic and subatomic level.

This overly simplified approach underscores one of the fundamental paradoxes of our current scientific reasoning which continues to try and squeeze together classical physics with quantum physics.[1] At present, the scientific community as a whole is largely critical of such a simplified explanation.

On the one hand Newtonian movements have greatly aided our perception of sidereal mechanics, yet on the other hand quantum physics has expanded our consciousness to conceive of a richer and deeper physical reality.[2] The combination of these two starkly contrasted perceptions has resulted in the search for a *different* and more unified perspective.

Let us consider this *different* avenue of thought and a different approach altogether. What if we completely remove the gravitational force and its hypothecated anchors out of the equation altogether? What if we begin by assuming that gravity doesn't exist at all. From here we can then consider that planets and stars are not bound to each other by some mysterious and ubiquitous force but rather by the same force which binds atoms and molecules together. Therefore, in this hypothetically assumed world, mass does not *need* to warp space and planets and stars do not need to be bound by their relative masses and densities, rather space acts upon space *within space* by the differentiation of harmonically defined quantum geometrical matrices produced by internal information.

---

[1] Philosophically the issue confronting our physicists is not the unity of a field but rather the underlying complexity of its nature governed by a form of superior intelligence.

[2] Distinctions of macrocosmic and microcosmic are subjective observations and have been conceived of as two fundamental states, when in truth Unity already presupposes our objective classification. Gravity, if it has been successful at anything, it has been at divorcing our reasoning from this Unity Consciousness.

Such a reasoning would expand the Kant-Laplacian model by the implementation of utilizing secondary (phenomenal) mechanics, such as accretion phases, to produce the finished product or even to induce its perpetuation as a potential energy source, after the harmonizing inducement is first established.

If we begin our investigations of early planetary and stellar formation processes from this assumption we are able to broaden our perceptions of physical reality. Subsequently from this broader context will arise a deeper understanding of life revealing an intimately woven quantum tapestry through which the fabric of life is derived.

Chapter Four

Perceptive Relativity

Electrostatic distribution ever seeks a harmonious equilibrium. It is an arbitrary and erroneous supposition which objects to this on the basis, that while atomic systems somehow maintain this balance, larger bodies do not.

Everything that we observe in our universe is derived from a principle of similitude and this is undeniably an existential quantum State. How then could we imagine that these so-called larger bodies, which emerged from such a quantum state, could possibly function by any different inherent laws?

Such larger objects, whether as planets or stars, operate by utilization of similar laws and principles. If this were not so the orderly arrangement of even our tiny solar system could not withstand. Consequently objections to the possibility of a large electrostatic equilibrium states is a fundamental flaw in the character of our egotism which narcissistically regards our stellar position as central, or paramount, to considerations of cosmic origins.

Conceptually this narcissism precludes a genuine pervasion of spatial geometry and harmonics and seeks to force an assumption into a tiny pigeon hole of causal reality thereby negating our 'relative' perception or our subjective basis of macrocosmic and microcosmic.

Reference frames, comparative to Einstein's Special and General theories, define these concepts of large and small by the act of the observer, who in the very act of observing negates the truth underlying the object.

The simple theory of Perceptive Relativity demands a radical departure from this narcissism and asks us to humbly consider that our conscious frame of reference is not the arbiter of Natures defining tone. Because of this it is necessary to facilitate an objective viewpoint that is consistent with nature's simple and fundamental laws. For as Richard Feynman once humbly admitted, "You recognize truth by its beauty and simplicity… When you get it right, it is obvious that it is right… truth always turns out to be simpler than you thought."

Chapter Five

Quantum Geometrics

"The building blocks of our theories are not particles but fields... continuous fluid like objects spread throughout space. The electric and magnetic fields are a familiar example... The objects that we call fundamental particles are not fundamental. Instead they are ripples of continuous fields". (David Tom, Professor of Theoretical Physics at the University of Cambridge, Scientific American, December 2012)

"Every ponderable atom is differentiated from a tenuous fluid filling all space merely by spinning motion; as a whirl of water in a calm lake. By being set in motion this fluid, the ether, becomes gross matter. (Nikolai Tesla)

The introduction of an initial quantum geometrical state is sufficient to provide the necessary momentum to begin a planetary formation process. From this position extraneous matter such as cosmic dust and gas are not needed though they play a definite role in further planetary development. Such an assumption partially derives from current Superstring theories, which suggests deeply embedded spatial fabrics and tiny circular strings vibrating at specific frequencies and intervals, while simultaneously producing spatial waves which in turn determine an element's "mass" system.

In my Special Earth Core Theory I propose that planetary

formations, such as the earth, begin *prior* to any collapse of nebular debris and any subsequent hedging in by gravitational constraints. As it is currently defined the Kant-Laplacian model fails to account for where the necessary interstellar clouds of gas and debris originate and is therefore unable to explain how stars can appear in locations without any pre-existent matter. If we consider the possibility that formation processes do not need pre-existent matter in order to produce phenomenal results, but rather inculcate innate spatial geometry and harmonic resonance, then all the prerequisites exist already hyperdimensionally. What the final "physical" result is that we observe, then, is nothing but a relatively perceived lower dimensional fixation, as an objectified frame of reference of the polarized medium.

Galactic formations would also follow the same quantum laws of attraction and repulsion with the only measurable difference being to scale. For example, when a mass such as the sun is produced it does not need pre-existent matter, rather the interior core geometry engenders it through the prefixing angles articulating harmonious waveform oscillations. The sun therefore does not need matter to collapse in order to produce energy rather it needs properly defined core geometry and spatial harmonics. The same logic and reasoning must also apply to the Earth[3] and other planetary bodies found throughout the cosmos.

---

[3] Earth was physically formed by a pre-existent and anterior quantum core geometry complementary to a tiny swirling vortex of space itself. Such a movement differentiated space electromagnetically thereby producing the secondary radially defined extensions of electromagnetic energy. Thus pre-existent geometry determines the extent to which the physical body will inform its surroundings by its mass by producing a moving

The driving mechanism behind interstellar formations is not restricted relativistically rather is induced by quantum geometrical matrices described phenomenally by frequencies in the electromagnetic spectrum. The fundamental laws of electromagnetic dynamics reveal the simplicity of this truth and show that life is indeed a byproduct of intelligence and information. Each permitted orbital shell in the Solar atmosphere has a specific frequency which determines *distance* and the possibility for a potential planetary formation defined by its specific frequency. Such an idea also necessitates that the existing planetary body must begin and end on a charge. In this sense each planetary body is coupled in their respective orbits by electromagnetic frequencies that define the Solar atmospheric zone.

These scientific parameters need further elucidation.

---

charge differential which can then be measured to harmonic logarithmic scales which have specific frequency intervals and a numerical correspondence. In this sense the earth atmosphere is a radially defined harmonic zone that extends from its core and each differentiation is a phenomenal measure of its massive charge. Therefore, temperature, mass or density, do not in any way distinguish radial distances of planets within the solar atmosphere anymore than they do in the earth's atmosphere, rather the harmonically induced moving charge is attracted and repulsed within each of these solar zones, i.e., *notes*. Each planetary orbit corresponds to a harmonic resonance defined by a radial distance.

Chapter Six

Mass and Equivalency

Mass is defined as the quantity of matter per unit of volume. Now since mass is derived from quantum geometrical states, and since volume is determined by the geometrical shape of its container, the conglomeration of atomic and molecular structures, which are prerequisites to mass, must be derived from a similar origin, therefore all physical mass can be considered as energy mass.

The law which converses gravity's dialogue is written $F \equiv G\, m_1 m_2/r^2$, where G is considered the gravitational constant and F is the 'Force' that one mass exerts on another Mass . Similarly the same law applies for electrostatic charges (made famous by Coloumb) and is written $F \equiv k \equiv 9.9^2/r^2$, where (k) is the electrostatic constant and where mass (m) is substituted for a point charge. Thus energy mass acts on another energy mass by the equivalent force made famous by Einstein. Both equations also hold similar inverse square laws that involve a property of interacting particles at a close proximity or in the act of moving away from each other.

In stating these elementary laws of force we imply that mass is nothing more than electromagnetically charged quantum geometry and that just because a mass is larger does not necessarily mean that it must have an increased action at a distance on nearby objects, thereby increasing its gravity. The reason for this is obvious, it is electromagnetic *charge* which produces the force acting upon another body. The physical size is therefore not an accurate indice to measure such a force (which is primarily centered at a core charge) whether we classify this as charge-density or not. The smaller objects, such as stars and planets, can have a much greater exerted force on another body simply because the internal core geometry predisposes the object to such a degree that it's charge density is greater and its overall attractiveness is increased. We can also state unequivocally that the harmonic zone that such a body finds itself in is a predetermined density and also a so-called *force*. Observationally then, the final physical mass we observe, such as the Earth, does not need to be a solid body to produce its active force since the quantum core geometry maintains a harmonic angular velocity conducive to a centralized sub-nebulated vortex.

By adopting the approach outlined here I suggest that the Earth and other celestial bodies do not need to be solid bodies, since the overall mass and density definitions are controlled by a sub-nebulated quantum core geometry. Thus the overall mass and density of the Earth does not need to be constrained by an outdated modus-operandi of an *Iron catastrophe* nor does its gravitational constant need to represent the force created as a result of it. Mass is not attracted by its size, nor its physical density, but by the geometrical acceleration into and out of the central vortex which is harmonically distinguished from other nearby acting bodies.

This explains why density is not directly associated with size because the issue is not a mass of particles acting on another mass of particles but the variant existential strength of the electromagnetic coupling within the continuing subnebulated core. The overall strength of the charge together with the harmonically induced angular momentum produce this force. This also explains why the acceleration of a body pulls inward toward its center at a specially defined radial distance while at the same time expelling objects at another. Both the centrifugal and centripetal forces are simply described in Coulomb's law for moving spherical charges.

All material substances are electromagnetic and possess an overall charge. There's absolutely no exception to this understanding that has ever been discovered.[4]

According to Newton's law of Force one discovers that this so-called *weight* cannot be directly related to a mass. For example, a body placed on the surface of the Moon will have a similar mass as it would on Earth, however, because the moon's core geometry and overall electrostatic charge acceleration is weak by comparison to the Earth, this body will not feel the same pull towards its center. In fact, the Moon contributes to a large extent in balancing the Earth's pull by a compendium of two opposite electrostatic charges in a state of general attraction and repulsion in orbital flux. The resistance of this body to free-fall inward is, according to Einstein, "a manifestation of all forms of energy in a body". Thus the weak principle of equivalence indicates that all forms of non gravitational energy must identically couple, but for this to hold true the ubiquitous force of gravity must also interact at a Quantum level, and to date this has not been

[4] Density, or weight, is a measure of the equal and opposite forces necessary to prevent downward acceleration of a body to fall inward toward the subnebulated proton mass core.

discovered or validated. In fact it is quite the opposite.

In no case whatsoever has quantum gravity ever been discovered yet the search for it continues unabated. It is in fact this persistence of a belief in gravity which is directly and indirectly the cause of a systemic paralysis in libertarian scientific thought. For example, as soon as we remove the need to justify gravity we essentially remove the larger mass obstacle standing in the way of scientific progress. We are also able to provide satisfactory answers as to why clocks run slower in the proximity of celestial bodies, why the special structure of physical objects cannot be explained by euclidean geometry but requires more sophisticated Reimanian geometry, and also why light rays do not travel in straight lines but are deflected by a body in motion. Each of these apparent anomalous gravitational constraints can be explained by simple electromagnetic properties derived from quantum core geometry and harmonic proportions with no need to include a magical and mysterious force —one that will *never be discovered.*

Inertial mass is a body's resistance to acceleration of another body and is a response to any form of external electromagnetic force acting upon this body. Both Newton and Einstein theorized that the larger the mass the greater the resistance would be to it being moved. Likewise both proposed that the g-force would be proportional to this mass so that the larger the planet the greater the tendency would be of a smaller body to fall into it. But since core mass cannot be quantified, nor substantiated, elemental mass such as iron and our ideas of inertia must be reexamined.

A body's resistance to the fall one feels is not the same for a planet or say a comet since the comet is merely the exploded debris of a former planetoid; the outer crustacean of a subnebulated proton mass core, therefore the combined effect of an accelerated proton mass and its quantum geometrical matrice must define its internal resistance, or to put it even more simply, it is the accelerated core charge of a moving magnet suspended in electromagnetically defined space which determines g-force.

In Einstein's general theory of relativity the physical consequence of the gravitational field is the warping of space by a mass of non-euclidean geometry and is used to define the curvature of such space.

According to his theory particles of light will travel along geodesics, or the shortest path within a four dimensional frame, however this euclidean frame merely provides the illusory appearance of light; an optical definition of a warped hyperdimensional reasoning ever seeking to validate the gravitational force at the expense of sound logic…time delays in radar pulses and Eddington hyperboles notwithstanding.

In view of my Special Earth Core Theory a planetary mass does not need to warp space and the only reason why 'bodies' fall at a specific acceleration rate is that hyper dimensional quantum space is being geometrically and harmonically accelerated at the Earth's core. This accelerated rate is 24 hours, or roughly 1,020 miles per hour, and is a perfected geometrical acceleration.

Einsteins view was that mechanical operations of

physically separate objects produced corollary effects on this space. Such a suggestion undermines the truth and places barriers between reality and perception. But even Einstein understood the need for further proof that validates such a premise, he concluded, "For a concept to make sense you need an operational definition of it, one that describes how you would observe the concept in operation". There has yet to be any operational validation of gravity as an independent Force despite numerous experiments using pendulums and time speed delays in light travel. In fact of all the known forces it is only gravity that has this claim to fame.

In all humility I would say that the jury is still deliberating upon gravity…

According to Coulomb's law an electric current must begin on a charge and end on a charge, therefore each planet or star in relationship to one another, would be considered as a point charge. And in this way the solar system as a whole carries an overall charge similar to an atomic system. Classification of solar shells is harmonic with a smooth progression; and planetary distances are determined by charge differentiation. According to Bode's law such a progression can be numerically quantified.

A dominant feature of this theory is its parallel with a fundamental note which corresponds to the longest wavelength of the solar shell, or its atmosphere, considered as a radially defined theoretical string which has an attendant positive and negative dipole. The theoretical mid-point would correspond to Jupiter and the overtone of the string would be a specific wavelength, or frequency, where each planet is a fraction of the fundamental note acting as a stationary node along the length of this electromagnetic dipole string. The resulting progression provides an approximate value with surprising accuracy.

The planetary chain within the solar atmosphere functions much like notes upon a string. Waves propagate along the electromagnetic, or harmonic, intervals between two points of charge.

Waves leaving the sun are like tipping over the first domino where the pulse travels along the entire string at specific velocities.

Waveform superimpositions exist between two binary stars as point charges in such a dipole system. Inter-locution between each charge is supraluminal and can be quantified as information by converting electrostatic harmonies consistent with entropic fluctuations, i.e., acceleration, into a set of specific radial ratios and geometrically defined proportions.

The conclusion is that stars are harmonic wave oscillators, possess an overall charge, and the solar atmosphere is a byproduct of such a understanding. Such a conclusion necessitates a reconsideration of not only how stars form but also how they further develop.

Our current relativistic planetary and stellar classification system is in need of serious revision.

Why?

Because, "no one has unambiguously observed material

falling into an embryonic star, which should be happening if the star is truly still forming. And no one has caught a molecular cloud in the act of collapsing. (Ivars Peterson, The Winds of Starburst, Science News, Vol 127, 30 June 1990, page 409).

"Precisely how a section of an interstellar cloud collapses gravitationally into a star —double or multiple star, or a solar system— is still a challenging theoretical problem… Astronomers have yet to find an interstellar cloud in the actual process of collapse. (Fred L. Whipple, The Mystery of Comets, Washington, D. See., Smithsonian Institution Press, Pages 211, 213).

Martin Harwit, writing in the Journal Science also observed that, "The Universe we see when we look out to its furthest horizons contains 100 billion galaxies. Each of these galaxies contains another hundred billion stars. That's 10^{22} stars all told. The silent embarrassment of modern astrophysics is that we do not know how even a single one of these Stars managed to form".

Harwit then goes on to list three formidable objections to all modern theories of star formation.

1. The Contracting gas clouds must radiate energy in order to continue their contraction, the potential energy that is liberated in this prestellar phase must be observable somehow, but we have yet to identify it.

2. The angular momentum that resides in typical interstellar clouds is many orders of magnitudes higher than the angular momentum we compute for the relatively slowly spinning young stars, where and how has the

protostar shed that angular momentum during contraction?

3. Interstellar clouds are permeated with magnetic fields that we believe to be effectively frozen to the contracting gas; as the gas cloud collapses to form a star, the magnetic field lines should be compressed ever closer together, giving rise to enormous magnetic fields, long before the collapse is completed. These fields would resist further collapse, preventing the formation of the expected star, yet we observe no evidence of strong fields, and the stars do form, apparently unaware of our theoretical difficulties.

Most of the anomalous features found in planetary and stellar formations arise from gravitational and thermal restrictions perpetuated by an outdated classification system and a flawed relativistic logic. Gravity does not lock planets in place around the Sun anymore than it pulls in cosmic gas and dust together into tight balls which are then ignited to become protostars. The clouds of gas and dust possess too much kinetic and potential energy, not to mention angular momentum, which would need to be removed first before any so-called matter could even begin to collapse. To date no evidence has emerged which conclusively reveals such a removal .

A similar logic applies to planetary formations. Contrary to popular opinion planets show no evidence of developing from mutual gravitational attraction of cosmic debris orbiting the Sun. This debris is much more likely to be scattered or expelled by the gravitational interaction. Experiments have disclosed that colliding particles fragment rather than stick together.

Gas is also dissipated quickly within space, especially the two lightest gases hydrogen and helium. Stars similar to our sun simply do not have enough orbiting hydrogen or helium to form even one gas giant let alone the four found in our solar system. So how do we suppose the formation of Jupiter, Saturn, Uranus, and Neptune? Might they be the result of some unknown feature of spatial geometry, the same feature which produces waveforms we erroneously classified as a heat source? Is it possible that each planetoid is really a harmonic frequency of a quantum geometrically defined solar core?

Chapter Seven

The Formation of Ring Systems and the Roche Limit

Another surprisingly anomalous feature of planetary formation processes are ring systems such as those observed around Saturn and Jupiter. The current opinion regarding the formation of these ring systems is a post remnant condensation disc.

Edward Roche, the 19th century French mathematician, was one of the first scientists to draw our attention to these planetary rings. According to his theory each planet possesses a critical radial distance parameter established by its core. This critical distance is the determining factor behind why certain materials such as those found in Saturn's rings do not gravitationally collapse into the planet while others do.

Recent observations, however, have shown the comet Hale-Bopp impacting Jupiter, which to the chagrin of many relativistic theologians was not captured; thus a mechanism must have existed that underlied the differentiation between the various materials, seeing how Saturn captures materials in its rings yet Jupiter does not.

Astrophysicists who have investigated this phenomenon have never considered it outside of gravitational constraints. From the very beginning their insistence is a further validation or substantiation of gravity never is it to actually explain such anomalies. The Roche limit defies gravity.

The particulate matter observed in these Rings should collapse into the planetary body or there should be some evidence of particulate dispersal by the plane delimited centrifugal force but such evidence is not observed. On the contrary all the evidence suggests that these particles are somehow trapped by an unknown physical property which has been unable to be traced back to an early planetary nebula.

The apparent challenge discovered in the Roche limit is an understanding of how and when the material making up planetary Rings reach their potential limit in radial definitions.

The particles trapped in these rings are trapped in the same way as planetoids are trapped in the sun's rings; by specific harmonic ratios emanating a sound wave oscillation from the solar core. It is the *keynote* of the debris which attracts it or repels it.

I contend that the Roche limit is not defined gravitationally but *harmonically* by electromagnetic radial extensions of quantum core geometry.

Chapter Eight

A Solution to the Angular Momentum Problem

This problem ultimately led to the untenable nature of the Kant-Laplacian nebular hypothesis.[5] The problem is, to put it simply, why the planets seem to possess most of the solar systems angular momentum while the sun possesses most of its mass.

The process by which an interstellar cloud of gas and dust is concentrated until it is finally held gravitationally to become a protostar is not known. The initial assumption is that the number of atoms pcm³ increases a thousandfold. However, gas has a tendency to disperse before the density becomes high enough for the force of gravitation to become effective. So what container is able to hold these atoms together and to continuously squeeze them? Any initial angular momentum causes excessively rapid rotations and contraction. So where is the mechanism that would be capable of gathering all these materials into a sufficiently small enough volume so as to make self gravitation effective?

The humble truth is that there is no reasonable astronomical scenario by which mineral grains and cosmic

---

[5] Again we must recognize scientific persistence in this regard, which is a tendency to justify Gravity, by beginning with an assumed gravitational foundation.

dust can condense. Most cosmologists now tend to point to missing mass and that somehow this dark matter and dark energy interact with the observable 4% of the material make up of stars and planets. The importance of such a conclusion cannot be overstated since it is largely believed that stars such as our sun radiate away their energy and subsequently their angular momentum. This perspective insists that this radiation is picked up by the planets, as *momentum*, allowing the sun to keep 99% of the system's mass while at the same time giving away its acceleration.

Such radiation is harmonic sound wave oscillations of space itself, not heat, not energy per say, but fluctuations in the waveforms which induce space, and it's interactive tendency with other planetary bodies within this harmonically defined space, to formulate heat and energy by such interactions.

We can easily dispose of the angular momentum problem by adopting the approach outlined here. Let us stop looking for excess mass, the issue is not missing mass, or momentum of a mass, but of quantum geometry and electromagnetically charged planetary and stellar cores, which is fundamentally independent of a prerequisite of mass or other so-called mass definitions given to it by gravity. Too many inconsistencies and problems arise by seeking to define planetary positions, and so on, by gravitational and temperature constraints. Let us begin with an altogether different assumption, one based upon the harmonic frequencies and quantum core geometry.

One objection to preexisting quantum geometrics deals specifically with how we currently classify stellar systems. This classification system suggests that stars are nothing more than oversized mechanical furnaces, therefore stars with a similar mass comparable to our Sun should have similar atmospheres that are slowly but steadily expanding into space, while larger stars would not show a similar tendency. Such a view is not based on relevant time-based observations but rather upon assumed stellar-mass classifications, stemming from their need to justify, by any means necessary, general relativity.

While the solar wind may in effect reduce angular momentum by their theory this is only in consideration of the accepted fact that stars are nothing more than chemically rich furnaces. This fails to account for one persistent perplexity: what perpetuates the energy in this furnace uninterrupted in time?

Energy from a star does not need to be preexistent as convertible forms of gas and dust rather quantum geometry moving within the physical medium polarizes space and engenders a unique hyper-dimensional bridge. Such a bridge is much like a series of hyperbolic geometrical lenses which produce inveterate waveforms that subsequently reverberate space producing the byproduct that we observe as photons and class distinctions within the electromagnetic spectrum. The Sun, and all stars, are geometrical lenses and their phasal developments[6] are not determined by burning up gases or fuel but by changes in the interior core geometry brought about by various phenomenally unknown mechanisms.[7]

---

[6] Phasal transitions are not that complex and the hypothetical fading of a star into and out of phenomenal existence does not need to be limited to momentum or mass.

[7] Current classification systems deal exclusively with temporal gravitational masses and temperatures. These parameters are in effect the culprit behind our observed anomalous characteristics.

Niels Bohr referred to this inconsistency between classical physics and quantum physics in his assessment made in 1958; he stated, "Indeed it became clear that the pictorial description of classical physical theories represent an idealization valid only for phenomenon in the analysis of which all actions involved are sufficiently large to permit the neglect of the quantum. While this condition is amply fulfilled in phenomena on the ordinary scale, we meet in experimental evidence concerning atomic particles with regularities of a novel type, incompatible with deterministic analysis. These quantities determine the peculiar stability and reactions of atomic systems, and are those ultimately responsible for the properties of matter on which our means of observation depend."

By removing these mechanical restraints we also remove most of our difficulties and let Schrodinger's cat out of the box. Current classification systems are based upon worn out deterministic views of the cosmos and are in need of serious revision.

Chapter Nine

Earth as a Spherical Conductor

Because the earth's core is a subnebulated proton/neutron mass its charge is predominately positive. This charge spreads uniformly over its body as homogeneous electromagnetic and atmospheric layers. The Earth removes its negative charge through the functionality of wave propagation. When the cores electrical charge discharges at the uppermost part of the mantle lid this facilitates the production of the electrical current around the equatorial band which is primarily situated in the tropical regions. The secondary action of this discharge is a magnetic dipole which combines with this electrical current to produce our atmospheric layers.

This is an overly simplified explanation of complicated atmospheric developmental processes but it does not seek to negate various chemical and thermal activities which take place in regional locations on Earth, rather it distinguishes formation processes atmospherically by revealing their origins electromagnetically.

It is easy to lose sight of the fact that the atmosphere on Earth is derived from the coupling of the electrical current in the magnetic field of the earth core. Therefore the atmosphere must be regarded as a secondary effect of a primary core charge and it's even distribution of spherical charge potential.

Both the Earth's electric and magnetic fields are perpendicular to the order of motion of this interior quantum core geometrical motion, thus both vectors of E and B are perpendicular to each other and formulate toroidal motions of the fields (magnetic and electric). The core charge momentum simultaneously produces centripetal and centrifugal actions considered as magnetic attraction and electric expansion forces, and also engenders the even distribution of the atmospheric layers. Subsequent movements of high and low pressure zones in the troposphere are a secondary effect of these electric and magnetic field motions.

While it is largely held that the earth's core is solid, geophysicists insist on speculating how a solid core could produce a magnetic field such as is observed on Earth. According to the Earth Core Dynamo theory, planetary development essentially ended with the Iron catastrophe, when this amazing development of an iron ball crashed into the core of the earth. Subsequent cooling processes than took place in the core releasing this converted heat energy to its surrounding mass causing it to melt in the process. Then through the transfer process of this perpetual source of heat, (begun 4.5 billion years ago), thermal convection settled into a steady state of transmission, which subsequently has continued unabated until the present; this despite the fact that the Earth on a daily basis emits enormous amounts of energy, which leaves us with two paradoxical anomalies in this context; iron entropy and magnetic polar reversals.

Chapter Ten

Metallic Entropy and Polar Reversals

Oxidation is a process of iron entropy. The iron catastrophe is an event which took place over four billion years ago. Now let's do the math here. We are being asked to accept, that with compressional rates as they currently stand, with electrical currents flowing through the iron core on a consistent and daily basis, with the repetitive periods of heating and cooling, with the addition of oxygen and water, that somehow this iron ball, with a frozen magnetic field, has continued unabated in a perpetual suspension of entropic decay blatantly defying all known natural laws which govern elemental substances? If such a gross defiance of natural law were to be accepted we might as well throw out all laws of thermodynamics together with our logical sense of deductive reasoning.

Here is the sad reality; the Earth Core Dynamo theory is failed logic and is unable to account for this anomaly as well as where the Earth acquires its perpetual source of energy, not to mention the well-documented polar reversals.

Magnetic pole reversals are a well-established fact of geophysical science. If the current dynamo theory is correct there should be no polar reversals. The fact that they exist is unmistakable evidence directing our attention to an unknown mechanism which operates upon the supposed *frozen* magnetic field in the earth core.

The Earth core functions similar to a battery and is able to conserve its charge in the core by an unknown process. It is quite difficult to conceive how a solid iron and nickel core could retain such a discharge continuously without any observable decline in its energy output or angular momentum.

Lithospheric thermal conditions indicate a consistent convection mechanism that is driven by the cores electrical current. And it is this current, along with the magnetic fields, which create atmospheric conditions and the various generated cells which surround the Earth.

The electrical conductivity produces chemically rich constitutions and climatic oscillations found within the Earth onion. The various internal shells simply provide an initial "charged" pathway for other charges to establish the radial vectors. For example, the continental or lithosphere crust, acts as a terminal discharge which produces an electrical field that is perpendicular to its surface. This electrical current then runs in a loop around the equator. Any potential shifts in the earth's angular momentum, or rotational velocity, must therefore be a special condition of the earth core and of this electromagnetic driving mechanism and any subsequent fluctuation in the core charge potential must have secondary effect upon the E and B vectors as well as the governing toroidal motions which provide fundamental defining constraints to atmospheric regulations.

Richard Feynman made this point clear when he said, "If a

magnetic field is suddenly changed it will produce tremendous electrical effects".

Current theory seeks to isolate the well-documented magnetic changes of the waning dipole field by simply ignoring the effect that this waning field is producing upon the electrical conductivity and atmospheric oscillations. Even a slight change to one dramatically impacts the other and a slight change in the Earth core will produce quantifiable effects such as the observed magnetic slipping of the dipole, and erratic weather patterns.

Chapter Eleven

Phasal Development and Electromagnetic Flux

The Earth as a *closed-circuit* conserves its charge quantum geometrically within the core. As long as this internal informing geometry maintains its consistency of "defining the space" the core charge remains unabated. It is a timeless principle that, "without attraction and repulsion there is no movement, without movement there is no life" (Viktor Schauberger).

The secret of life is polarity and without opposite poles there is no attraction or repulsion. The two natural modes of this opposition are hyperbolic spiral motions into and out of the earth core defined as the excentric centripetal force and the concentric centripetal force. The undulating spiral currents act as waveforms, and are induced from this central position; appearing as eddy currents would in a whirlpool. These currents *involute* in the north and *evolute* in the south extending inward and outward by radially defined vectors; a simultaneous movement of centripetal and centrifugal field aligned currents.

It is these governing electromagnetic principles which determine the seasonal oscillations occurring in the atmospheric encasing of the Earth. Every phasal transition period on Earth is dominated by a specific action upon its tilt, the angle of its obliquity, and the axial and orbital velocity rates. These changes are then transferred to seasonal variations both annually and precessionally, producing enormous consequences in torsional balance, and potential alterations in both the magnitude and direction of wave propagation. Such changes act upon the buoyancy and the contraction of the lithospheric crustal layers and the oceanic temperatures which follow the same toroidal motions twisting the Earth onion.

Richard Feynman made this point abundantly clear, he stated, "Suppose the magnetic field were to disappear… there would be a changing magnetic field which would produce an electric field. If this electric field tries to go away the changing electric field would create a magnetic field back again. So by a perpetual interplay —by the switching back and forth from one field to another— they must go on forever. It is impossible for them to disappear. They maintain themselves in a kind of dance— one making the other, the second making the first propagating onward through space".

Chapter Twelve

Wave Propagation and Seismic Activities

Over 90% of the Earth's interior is inaccessible to direct observation and is therefore unknown. Geophysicists study rock fragments called xenoliths, which are volcanically brought up from depths, to provide examples of the upper mantle. However most of these samples originate at depths no greater than 150 to 200 kilometers. Thus investigators are forced to use indirect geophysical approaches such as wave oscillations and seismic activity to formulate scientific analysis about the Earth's interior.

The interior of the Earth produces two primary forms of wave oscillation. These are classified as P and S waves, or Compressional and Sheer waves. Compressional waves are identical to sound waves in that the deformation is in the same direction in which the waves are propagating. On the other hand, Sheer waves are perpendicular to this propagation and seem only to be transmitted through solids, since by definition fluids do not possess enough rigidity that would enable them to support the transverse deformation that is generally associated with Sheer waves.

Understanding wave propagation is paramount to an accurate portrayal of current density and pressure indices utilized by geophysicists. These indices serve to validate current seismological models such as the one presented by Backus and Gilbert. In their model body waves are inverted and data concerning free oscillations is recorded accurately. Thus density in velocity profiles are compiled through the observation of the Earth's surface which enables geophysicists to make estimates concerning the internal composition of the earth.

Determinations concerning interior density distribution (Vp and Vs profiles) provide very flimsy reports of elasticity at any level lower than 150 to 200 kilometers. If there was consistency seismic waves would appear to travel in straight lines. The fact that they do not is provisional evidence supporting an aspect of torsional wave propagation originating beneath the outer core. This rotation must to some extent coincide with the apparent observational direction of these waves. Current models show a marked deflection of arc with a noticeable *inability of body waves and free oscillations to penetrate the core.*

P and S waves are analogous to light and sound waves which demonstrate variations eerily similar to optical processes of reflection and refraction. Studies of refractive indices have indicated the Earth's body wave demonstrates a similitude to known references, tracing paths considerably comparable to known torsional or spiral motions. Comparisons of these motions with known refractive and reflective data has caused many geophysicists to consider the Earth in the capacity of an enlarged distortion lense.

The paths of both the P and S waves demonstrate that the Earth core, "is dancing to its own beat, spinning measurably faster than the rest of the planet", (Science News, Vol 154, 25 July 1998, p.58) and that the Earth core, or some other rotational aspect of its external components, are moving relative to one another, following an angular velocity which to some extent coincides with a spiraling motion. This motion could be the cause of these internal variations that seem to indicate *distinctive* definitions of elasticity and density with greater depth.

Current relationship parameters between seismic wave velocities and internal elastic properties of the Earth is given by the equations $Vp= \sqrt{ks+4/3\mu/p}$ and $Vs=\sqrt{\mu/p}$. Both P and S wave definitions are dependent upon the differentiation of rigidity and density classifications.

Incompressibility (Ks) provides velocity measures while for fluids rigidity equals 0, so that $Vp=\sqrt{Ks/p}$ and $Vs=0$. Because the Earth is anisotropic and does not possess invariant elasticity or rigidity it requires a modification of these equations to include the directional dependence of velocities that are noticed.

According to the currently accepted 'seismological Earth model' P and S waves are measured within the framework of the solid earth core. In fact, it has been almost 100 years since seismologists have made any additional determinations aside from this preconceived mechanistic constraint.

In the mid-1930s British geophysicist Harold Jeffreys and the New Zealand seismologist Keith E. Bullen collaborated to produce travel timetables and velocity depth profiles. These became known as the Jeffrey-Bullen model and have since been utilized as the standard by which P and S waves are analyzed. The fact that this model has remained unchallenged for so long does not stem from its veracity but rather from the immobility of other scientists to discover or put forward any other model which is able to account for and explain the noticeable inconsistencies and problems associated with it.

The fact remains that body wave seismology has many problems. One example is its inability to measure velocity of P and S waves within this established framework of a solid earth core. It is also unable to provide accurate data of interior elasticity or rigidity below a certain continuity threshold. A number of these thresholds display inconsistent density which are invariant and do not coincide with prescribed density and pressure standards coincident with these depths. Essentially the measured wave propagation's lack uniformity and reveal inconsistencies within the accepted framework of density and pressure models that one should see if the Earth core were solid.

For a solid earth core to remain tenable density rates that are changing with depth, as one travels from the Earth's surface into the core, should match with known pressure gradients and show marked comparison with solid structures as can be shown by the incompressibility factor. However, this is not what one observes, on the contrary density does not in any way coincide with known or accepted velocity rates for pressure gradients which should be the case if the Earth were a solid mass. Numerous areas show interior distortion patterns inconsistent with the solid Earth, acting in many instances as though the Earth were indeed some kind of distortion lense.

In my Special Earth Core Theory (S.E.C.T) I propose that

this wave propagation variance can be accounted for by a proper consideration of electromagnetic oscillations emerging from a 'continuing subnebulated core', as well as its demonstrated torsional twisting action that is transferred to the upper mantle lid and lithosphere. Instead of viewing 'fault' activities as being due to the rigid movement of tectonic plates I am suggesting that these measured body waves are in fact expansive waveform deformities and the combined effect of the toroidal movements, that over a series of time twist the elastic skin of the Earth, producing fractures and hot spots of magna plumes. In this sense wave propagation is viewed at a point of origin within the Earth core and for this reason no other secondary reverberations due to these fractures are able to penetrate there.

Seismic activity can be excused as a reciprocal effect of electromagnetic oscillation which is subsequently carried to the Earth's surface. Over considerable amounts of time the energy created in this twisting action induces forms of crustal displacement and the slipping that follows, but not penetrating the core due to its velocity and interior motions.

The fracturing of crustal layers, and the so-called fault lines, indicate areas where the pressure is being *relieved* and where such a pent-up energy is being dispersed. Seismic activities are therefore nothing more than combined frictional pressures and dismissive electromagnetic wave velocities.

Adopting such a view provides a viable alternative to the standard model. It is simple and is able to explain the inability of fracturing reverberations to penetrate the Earth core and is a clear proof of the cores velocity and the fact that its electromagnetic radiative transfer rate is greater than the encasings which surround it. A similar process is found above ground where certain radio waves cannot penetrate the ionosphere due to its greater frequency. This simple explanation is sufficient to account for this anomaly.

Inferential measurements of free oscillations and P and S waves reveal motions, to or away from the earth core, that are insufficient to account for the observed reflective and refractive indices by acceptance of a solid earth core. Reflectivity can only be caused by a large increase in seismic wave propagation velocity of certain depths, yet these depths should possess noticeable increase in pressure and density that act to decrease velocity, however the opposite is true.

It is generally admitted that wave velocities generated by surface seismic activity do not provide accurate density and elasticity information. Despite this fact the current model continues to be used to explain these pressure and density differences based upon superficial calculations of observed periodicities in the oscillations. Further analysis of this information subsequently has led to such erroneous conclusions regarding the Earth's interior.

A reformulation of this analysis should begin with the premise that all known wave oscillations originate within the Earth core, thus any secondary, or tertiary, seismic activities should be able to be traced back to their point of origin. With such an about-face we remove the necessity to make assumptions regarding absolute plate rigidity, magma hotspot fixidity, and mantle homogeneity energies that are required for the current plate tectonic model.

S.E.C.T proposes an internal core torsional rotation. As the subnebulated electromagnetic core rotates it transfers its wave-forms to the crustal layers by an independent and increased velocity of motion so that the various layers, or shells, rotate out of time with the core. Such an insynchronous movement produces torsional imbalance and friction between these various layers. This would explain why observed continental movements are relative to the rotational axis of the earth. Polar wandering is therefore not a direct result of a change in the Earth's magnetic field, it is the combined effect of some sort of mechanism which induces continental drift convecting crustal layers into and out of the Earth's interior.

In order to demonstrate this we can make comparisons between the magnetic poles in the observable auroral ovals. These ovals reveal the magnetic convergence of field lines. Now, if we suppose that the internal core torsion is the cause of such field lines it is obvious then that these field lines would follow this actual rotation. And as they do the Earth's crustal layers are also being induced by a similar internal engine "to move" which ends up appearing to alter the geographic pole. This explains why the frozen magnetic moments in xenoliths are not the direct result of a change in the Earth's magnetic field but of a secondary mechanism caused by this internal core torsion. There is also reason to believe that crustal displacements coincide with these polar magnetic reversals. Why?

Crustal displacements appear to be consistent with time and to coincide with electromagnetic wave propagation.

We must begin to view the Earth system as a single entity and not isolate one aspect of its system to explain others.

In order to accomplish this it is necessary to have a proper understanding of Earth core physics.

Any determinations of interior density profiles must arise from observable and consistent facts, which can be easily convertible into other features of the Earth's system.

Present scientific studies continue to utilize the outdated Lehman hypothesis first proposed in 1936. I believe it is time to consider a radical alternative, one that will not create, but rather explain these anomalous physical features of Earth physics. If we can simplify our current oscillation models by adopting the premise that wave propagation originates in the earth core we obviate the need for extraneous assumptions about the deep mantle and core. We also find that numerous observed paradoxes with seismic and volcanic activity can be easily explained.

It is common knowledge that during earthquakes, (which are seen to occur during the evening), pyrotechnical displays of luminal protuberances emerge out of fault fractures. At the same time the Earth appears to vibrate, producing a tonal variance which can last from 54 minutes, up to two days. The fact that these two 'effects' of the internal core torsion are consistent with electromagnetic phenomena is indirect evidence of an active physical concurrence between clearly demonstrated helioseismology and geoseismology.

Both the Sun and Earth appear to ring as a bell acting as if they are both *hollow*.

In every instance we observe phenomena which in one way or another coincides with an electromagnetic phenomena. Seismology is a secondary study of a primary source of origin for wave propagation. Free oscillations also provide us with a means of inverting calculations for the periodicity of these oscillation data and simultaneously provide a detailed perspective of material constituency for accurate density profiles but not below the 200 kilometers mark.

In every known instance averages of this density of material constituency utilizes invalid approximations for further depth. In fact many of the observed densities of material constituencies do not concur with these known averages. For example, we observe areas of discontinuity where velocity increases or decreases, such as the known decrease observed in the tectonically active region of the western United States. Here Vp and Vs decreases over a limited depth range before resuming their usual increase with depth. Sharp discontinuities are also observed at depths of 400 to 670 kilometers where the upper and lower mantle separate. This so-called transition zone shows an anomalous rapid increase in wave velocity with depth, as compared with the rest of the mantle. With the present solid earth core theory these anomalies cannot be accounted for.

Continental disparities that we observe are the direct result of torsional momentum as it feeds into a torsional well at the North Pole. This coincides with the observable ellipsoid and polar flattening. If these motions, and their secondary effects, were taken seriously it would be sufficient to account for noticeable discontinuities with depth.

Hydrostatic pressure in rocks and various minerals form from electrical conductivity and rotational friction causing as a by-product the formation of *water*. As this twisting action becomes more and more pronounced it causes increased temperatures thereby increasing the likelihood of water creation. The water will ultimately flow into low pressure zones between the upper and lower mantle. Here the water could collect into estuaries or aqueducts becoming a means of crustal lubrication between the lithosphere and the asthenosphere. This would be sufficient to account for the discontinuities and variations in crustal thickness that one observes throughout the Earth.

Water is highly conductive relative to rocks, therefore, if water were found between the upper and lower mantle in the transition zone, we would see a definite increase in velocity of wave propagation in this area. And this is precisely what is observed. One also notices a low-velocity zone between the 80-400 km depth, which does not coincide with the Preliminary Reference Earth Model, (PREM), or its even density distribution.[8]

---

[8] These density parameters derive primarily from arbitrary classification systems based upon the gravitational constant. If we are to make any significant scientific progress I believe we need to remove these gravitational constraints and not make assumptions based upon unprovable calculations of gravitational acceleration rates.

Chapter Thirteen

Polar Flattening and the Earth Ellipsoid

There is much speculation regarding the extent to which gravity plays a role in the flattening of the Earth poles. In fact theoretical models dealing with the formation of a mass fluid with hydrostatic equilibrium, and a comparable rate of rotation of the Earth, have shown lateral density anomalies inconsistent within gravity parameters. Mean estimates of the Earth's polar flattening are about 1/298.257 compared to a mass model of 1/299.5 .

In S.E.C.T I propose that lateral heterogeneity found in density distribution is a direct result of the continuance of sub-nebulated torsional motion of the electromagnetic wave propagation emanating from the Earth's core. It is such a geometrically defined torsional motion which alters the physical encasing of the Earth.

In the conventional sense the 'Alpha and Omega' effect are directly associated with the centripetal contraction into the Northern polar torsional well and the centrifugal expansion which is equatorially defined. The amplification produces intersecting differential lines of electromagnetic force which depend on the core rotation. Such field lines wrap around the earth and their subsequent fluctuation causes fractures, faults, subduction zones, and hot spots of magma plumes. In effect, there is an intimate connection between this core geometry, the polar flattening, (the

Alpha effect), the toroidal loop that follows an East-West electrical current (the Omega effect) and the lithospheric encasing.

Magnetic contraction causes polar flattening and the electrical buoyancy, equatorially plane delimited, causes the 'bulge'. Thus the irregularity observed by geodetic satellites in the geoid is not a secondary effect of gravity.

In some localities irregularity is up to 100 meters higher than what ideal reference frames suggest for the Earth ellipsoid and in others up to 100 meters below it. Other gravitational anomalies, such as larger masses of mountain ranges possessing less gravity density than are supported by current standards, or why the oceans possess more density than gravity profiles, indicate an incongruity between these gravity density profile standards and these measurable observations.

Chapter Fourteen

Secular Variation and Magnetic Anomalies

Recent observations of the Westward drift of the magnetic field have shown evidence of secular variation.[9] These observations have been made since 1540 and have revealed that the direction of the magnetic field in London has nearly completed a full cycle with a peak amplitude of 70°. The superimposition of these observed changes together with other observed changes has resulted in the formulation of a magnetic map grid system.

Current Earth models begin with the persistent belief in a solid core and regard secular variation as a secondary effect caused by the dynamo mechanism. This, despite the fact that the variation is on such a short time scale that makes it inconsistent with the dynamo theory as it is currently understood. Because the Earth core is believed to be solid the implication is that such a variation must arise from another source such as the outer regions of the liquid core. Their reasoning is that if the source were any deeper the variation would be attenuated by the electrical

---

[9] The observed displacement of the field has continued to follow a westward course through time at a rate of 0.18° per year. Since 1850 the strength of the dipole component has decreased roughly from 8.5×10^{22} to 8.0×10^{22} amps/m² showing a waning field. The origin of this drift and decrease in strength is not known but it may be possible to explain this based upon the parameters setup in S.E.C.T.

properties of iron's conductivity making variation detectability impossible at the Earth's surface.

In my opinion the Westward drift provides evidence in support of my Special Earth Core Theory which, according to this theory, the variation is onset by changes in quantum core geometry which could be induced by external forces such as an orbital position change of the solar body in the electromagnetic field of its annual or seasonal space. Because the field potential has been transitioning so quickly over the past four hundred years it has had an indirect and direct impact on the internal electromagnetic torsional balance of the Earth.

The westward drift provides further evidence of how the Earth core generates the main field. Variations in the velocity and directional axis of the core also demonstrate a winding-up phase which appears to move backward relative the general rotation before it begins to unwind, thus the magnetic field appears to move Westward. Secular variation of the magnetic dipole coincides with this axial rotation of the core which to some extent is susceptible to external forcing and a theoretical change in its axial direction. If a change to the field gradient of the solar atmosphere was strong enough it could potentially cause a rapid fluctuation of this internal torsional angular velocity thereby inducing a geographical shift of the Earth's main field and lithospheric continuity.

Chapter Fifteen

Auroral Ovals

Auroral ovals are additional evidence supporting the torsional rotation of the magnetic field generated by the Earth core. These ovals display a tendency to circumnavigate a geographical depression or what I refer to as the Northern Polar torsional well. This depression of the ellipsoid is caused by the inner hyperbolic concavity of the core which comes as a direct result of the cores spinning motion and its propensity for the magnetic field to circle into the core as water circles down the drain. The concavity itself is the geographic compressional effect engendered by the magnetic phenomenon producing 'ovals' as it encircles this depression.

As the magnetic moment of the earth fluctuates in time its orbital motion traces a path coincident with these ovals and defines a Precessional characteristic. Such periods are marked out on the constellation of characters of the celestial sphere. Thus the celestial sphere seems to rotate in tandem with the northern circumpolar Stars. Through each Precessionally defined "Age" seven pole stars appear to define this passage as well as the concurrent electromagnetic activity at specific intervals. For example, because the Earth core rotates slightly faster than other parts of the earth, over these vast periods of time the crust begins to displace its plane of balance in these motions and no longer coincides with the hyperbolic torsion. As a result of this imbalance the angle of the ecliptic begins to shift appearing to fluctuate between 21° to 24° with respect to the Earth's orbital plane. Such a shifting causes an imbalance in weight prescription resulting in a necessity to realign its proper even distribution of movements.

This ecliptic shifting explains the Earth's wobble to be the result of accruing imbalances of independent motions between the core and the rest of the earth. Taken by itself this Precessional orbital flux is insufficient to account for the noticeable periodicity of a waning field and overall crustal displacement. Thus we must consider that the perihelion and aphelion axis of these auroral ovals are secondary effects of this cyclical activity.

Chapter Sixteen

Tectonic Plate Revolution

"...notions of fault revolution that have long been accepted are now being called into question. Traditional avenues of research have lost their potential to yield surprising insights." (The 95th Dahlem Workshop)

In more recent years it has been posited that the Earth core produces hydrodynamic waves. As the core rotates it rotates at a similar velocity to the mantle but with a slower wave propagation due to the medium through which the waves propagate. There is thus a slight decrease in velocity of the Earth as these waves propagate radially. This has led investigators to conclude that the core must be moving faster than the outer core. Recent evidence has supported this theory.

In August of 2005 a team of geophysicists announced in the journal "Science", that, according to their estimates, the Earth's inner core rotates approximately "0.3 to 0.5 degrees faster" per year then the observed rotation rate found at the Earth's surface.

While these rotation velocity differences are only slight, over significant amounts of time torsional imbalance and uneven weight distribution tends to accrue. The rapid onset of a dipole dispersal, coupled with inequality in the steady-state weight distribution, contributes to crustal displacement.

The tendency for a system to find balance necessitates a redistribution not only of the E→ and B→, but also of geothermal conditions, oceanic positioning, as well as crustal emplacement and mountain ranges. This leads to the creation of new fracture or fault zones over new Upper Crust layers and the lower subducted areas.

Such a rapid movement produces an enormous friction melting Lower Rock layers and producing hot plumes which can eventually lead to volcanic hot spots.

The general theory of plate tectonics is isolated and heavily constrained and is unable to account for any of the observed anomaly features of the geoid, such as the exclusion of incipient plate boundaries, volcanic chains,[10] or any fertility variations or other melting anomalies.

S.E.C.T expands our thinking regarding continental mobility and island fixidity.

The tectonic plate theory has created a number of observational anomalies that are difficult to reconcile. Current theory views these rigid plates as possessing very narrow boundaries that overlap a more chemically homogenous shallow mantle with simple radial temperature gradients. Unfortunately this does not account for the more broadly defined regions of the crust, other aspects of continental rheology, or so-called melting anomalies, or hotspots found along the ridges. Nor does it

---

[10] Volcanic islands have generally been viewed as the tops of narrow hot upwellings from somewhere deep in the mantle. The plume hypothesis came as a direct result of this views inability to account for volcanic features that plate tectonics cannot explain.

account for volcanoes that have been found far away from plate boundaries.

No agreement has been forthcoming as to what the driving mechanism for plate tectonics even is.

Such anomalous features have not gone unnoticed. According to geophysicist Don I. Anderson of Caltech University, there are two primary approaches for potentially reconciling the geological, geophysical, and geochemical anomalies associated with plate tectonic theory. His opinion was that we need "additional features to the basic theory," which includes "something outside the framework of plate tectonics." At the same time he says we need to drop the assumption "that plates are fixed, rigid, and elastic, and that the underlying mantle is homogeneous."

In S.E.C.T hotspots do not need to be isolated at plate boundaries nor do they need to originate deep within the mantle. These hot spots are the result of rapid torsional movements brought about by the overall imbalance in the even distribution of angular velocity and weight.

Anderson also concluded that, "although the plate tectonic theory has great exploratory and predictive power it seems to fail in regions of distributed continental deformation, continental break up, large igneous provinces and island chains. Geochemical models of so-called mid-plate volcanoes, and large igneous provinces have evolved independently of plate tectonics". This necessitated the popular plume hypothesis and is why attention has been drawn "away from the true source of this phenomena".

Both the Pacific and Atlantic mid-oceanic ridges are focal

points of geographical exhumations for the interior wave propagation. As the upper crustal layers are being slowly twisted and simultaneously moving toward the northern regions enormous amounts of pressure is building.

There is abundant evidence showing that crustal displacement is somehow connected with historical climate change. For example, geologic evidence shows tropical *fauna* in Antarctica and of glaciers in areas of low latitudes such as in Africa. These anomalous characteristics have yet to be explained by current theory.

If we consider the parameter set up by S.E.C.T we find a viable explanation; crustal layers move in general synchronicity over the upper mantle into the northern hemisphere. Because of incongruous velocities of these layers the velocity mass slowly moves out of sync with the core. When this imbalance becomes acute there is great disturbances on the face of the Earth. New lands arise around the equatorial regions and the oceans are forced into the polar areas.

When eventual balance is restored the oceans return to the equatorial regions by means of the Coriolis effect and the submerged polar areas once again become fit for habitation.[11]

As the magnetic pole follows the path of the auroral ovals it alters the geographic position[12] of magnetic north.

---

[11] Currently we are in the middle to end stages of a receding magnetic dipole which will in all likelihood coincide with the polar areas arising out of the cold frost and back into more temperate regions. The implication derived from this statement is that the melting ice caps are being caused by secular variation.
[12] Evidence of the last shift of the Earth's crust most likely took

Interestingly Albert Einstein in a forward found in professor Charles Hapgood's book, "Earth's Shifting Crust", drew a similar conclusion, he said, "A great many empirical data indicate that at each point on the Earth's surface that has been carefully studied, many climate changes have taken place, apparently quite suddenly". Einstein then went on to clarify what he believed may have caused such noticeable and rapid climate changes– crustal displacements and ice deposition.[13] He said, "Such displacements may take place as a consequence of comparatively slight forces exerted on the crust, derived from the Earth's momentum of rotation, which in turn will tend to alter the axis of rotation of the Earth's crust…In a polar region there is a continual deposition of ice, which is not symmetrically distributed about the pole. The Earth's rotation acts on these unsymmetrically deposited masses, and produces centrifugal momentum that is transmitted to the rigid crust of the earth. The constantly increasing centrifugal momentum produced in this way will, when it has reached a certain point, produce a movement of the Earth's crust over the rest of the Earth's body, and this will displace the polar regions toward the equator."

---

place in recent time, at the close of the last Ice Age, where the magnetic poles were situated in the Hudson Bay; 60° North latitude and 83° West longitude.

[13] The present melting of the ice caps is a secondary effect of secular variation and orbital eccentricity in the angle of the ecliptic.

Chapter Seventeen

Magnetic Anomalies

It is well understood that the Earth's magnetic field has not always been oriented as it is today. On average it is found that the dipole component reverses every 300,000 to 1 million years.[14] On geologic time scales, however, the actual reversals are quite sudden taking place in a 6,000 year period.

Opinions tend to diverge when consideration is given to the accurate compilations of periodical isolation data. Without a general consensus of time elapsed data to rely upon many scientists are forced to lean upon speculation which brings no relevancy to the discussion.

In fact after viewing all the available evidence in his time Professor Hapgood determined that there was evidence of at least three to four displacements coinciding with secular variation in the last 100,000 years. Hapgood stated that at the end of the last Ice Age, in roughly 11,500 BC, the lithosphere showed evidence of slippage some 30°. The onset of this movement was so rapid that he believed it was missed altogether by paleontologists. His argument was that isostatic balancing was so greatly challenged during this time, and it happened so quickly, that any noticeable demonstration of crustal displacement escaped modern detection.

---

[14] These are the most conservative estimates and not all scientists are in agreement with them.

Hapgood places the last three crustal displacements in the Pleistocene era and directly associates them with climatic upheaval. The evidence is certainly suggestive, in light of S.E.C.T, that the sudden onset of Ice Ages, extinction level events, and magnetic discrepancies all seem to be related. Current theory consistently seeks to isolate each phenomenon into specialized gloves that only certain hands are able to fit into. This produces unnecessary difficulties for cross-disciplinary field comparisons. This, coupled together with staunch ideological entrenchment, places independent thinkers in the position no better than Copernicus or Galileo. In order to remove these roadblocks to future scientific progress I believe it is necessary to radicalize specialized thinking and promote parameters of free thought unconstrained by topological fear of career suffrage.

All the available evidence is to my mind highly suggestive of periodical magnetic field reversals that follow internal core geometry and extraplanetary scalar field changes. If current theory regarding the eventual dissolution of the dipole is correct we have an indirect cause of current climatic oscillations. This harmonizes with known and well established electromagnetic dynamics.

Now, for the magnetic field to be changing means that the ionosphere is also fluctuating, which inversely supports my contention that core dynamics are involved. The reasoning here is simple with no need to make extraneous suppositions.

The conclusion is that with the dipole component waning it is altering the geobiospheric and climatic balance.

Chapter Eighteen

The Causation of Atmospheric Temperature

The orbit of the Earth around the Sun is an ellipse with an apogee of 1.47 x 10 (8th power) kms in January and a perigree of 1.52 x 10 to the 8th power kilometers for July agreeing with the two solstice periods. This simple orbital rotation period (coupled with lunar librations) produces temperature and climatic oscillations which remain relatively consistent over time.

The Earth rotates on its axis with an angle of 23° 30' with respect to the plane of its orbit. As a result of this tilt it is contended that sunshine is more direct at a given latitude during summer compared to how it is during winter in either the Northern or Southern hemispheres.

Poleward of latitude 66 ° 30' the tilt of the earth is such that, for at least one complete day, and as long as six months at 90°, the sun is above the horizon during summer and below the horizon during winter. This demarcation is self evident, however, what is not so evident is the manner in which solar heating is symmetrically distributed.

This basic description is often used to explain the manner in which the sun heats the Earth but is very misleading. What the sun actually does is produce various frequencies of wave propagation by geometrically polarizing space. As the sun rotates it whips up electromagnetic space producing harmonic oscillations which undulate in spirals. Because the Earth is generally charged a large part of these charges are collected by the Earth and those that are not are deflected by the magnetosphere. Also a large part of these wave particles are photons, which when interacting with the magnetosphere and the Earth's electrical current, causes an excitation of the gas molecules within the atmosphere. Such a dynamic interaction is known as the photoelectric effect and is where electrons are freed up producing the sensation of heat in the transference process of ionization.

Aside from the interaction of solar wave propagation the Earth's own diurnal component caused by its rotational velocity produces evenly distributed thermal conditions. Heat, in the general sense of the word, is not traveling through space, it is produced predominantly in the earth core and by friction of wave interaction and particle acceleration through ionization of the gas in the atmospheric layers of the earth onion.

Photons do not carry heat.

It is very misleading to assume that space-time distortions will produce heat on another body. If that body has a weak electromagnetic charge chances are it is not heated as much as commonly thought.

Photoelectricity produces a measure of heat in the transfer of energy states and is a primary source of lower tropospheric radiative thermal conditions. Due to specific electromagnetic and gaseous properties on the grid, the Earth is subject to variations of seasonal vicissitudes and temperature regulations, however, in all cases, heat *rises* it does not *descend*.

Angular photonic interference is simply used to explain daily and annual temperature variance. This interference is divided conceptually along imaginary lines of latitude and longitude, so that around the equatorial regions temperatures are consistently higher, while as one travels towards the poles temperatures are much cooler. Typical definitions of temperature profiles are determined by Vertical incidence and the production of ionization per unit of volume, more than say slant entry along the higher latitudes.

Definitions of this kind are highly misleading. Temperature variations are not due to entry of solar radiation. Temperature oscillations around the equatorial regions have less to do with the angle of energy entrance then they do with the Earth's electrical current, likewise temperature and climate variations moving toward the poles have more to do with the Earth's magnetic field than it does with the angle of entry. In the process of energy transference heat is produced more predominantly in regions of the earth more heavily laced with electrical fluids. As one travels perpendicular to the equator, or away from the equator, photo-electricity becomes more predominant on the surface rather than in the current as more forms of radiation waves interact with the lower atmosphere. The equatorial regions insulate lower tropospheric heat trapping it beneath the ionosphere. Finally as one reaches the negative and positive antipodes this radiation is deflected toward the equator and minimal heat is produced.

A great majority then, of Earth's thermal conditions, are generated by the core. Equatorially the wave propagation is electrical and photonic interaction causes thick and viscous subluminal oven type temperatures. Because heat ascends, and because photons carry no heat, sensation of heat in these equatorial regions is not derived from solar incidence.

It is well known that in the winter the sun is actually closer to the Earth than in summer so how is it then that distance parameters determine heat or temperature seasonal variations? The magnetic poles have direct solar radiation for six months straight yet have no significant temperature increases? Around the equator it is hot year round and this despite the angle at which solar radiation impacts the earth? Thus it can be easily ascertained that temperature regulation and climate change have more to do with Earth core dynamics than with the angle of photonic entrance.

Chapter Nineteen

Ice Ages and the Melting Ice Caps

According to S.E.C.T the polar ice caps are melting due to the observable secular variation and the westward drift in direct connection with the ecliptic angle.

Due to the polarization of the Earth's crust by the magnetic field it has been noticed that the poles have changed position, relative to the surface, at least 200 times since geological history began over 100,000 years ago.[15]

The pole seems to have moved to its present position in the middle of the Arctic Ocean gradually beginning around 25,000 years ago and finally settling into its present position around 13,000 years ago.

Since this time the Earth has been in a relatively calm interstadial period. On the other hand the later phases of the Paleolithic Age (15,000 B.C to 10,000 B.C) were characterized by erratic climate conditions as the ice caps over Europe retreated. This generated rainfall which allowed vegetation to revive in the more parched areas such as in Africa. In other areas ensuing droughts forced mass migrations of man and animals to more fertile areas. As of yet we have no accurate understanding of what the source of this great and sudden change in climate was.

---

[15] The last magnetic pole was situated in the Hudson Bay latitude 60° north and 83° west longitude.

There have been various suggestions, among others, that the inclination of the Earth's axis was greater, or that a submersion of the continents under water might have produced a marked decrease in temperature.

Interestingly the glacial age produced ice on a stupendous scale which could imply that it was preceded by heat on a stupendous scale. The sudden nature of some of these climate changes are further evidenced when we observe the extinction patterns in the last Ice Age. Here animals were frozen dead in their tracks, some in the very act of eating. Others were swept up by the hundreds of thousands (in herds) and frozen into the Upper Crust of sedimentary deposits of what was to become the frozen tundra. At the same time we discover in western North America no evidence of a natural drift deposit in the drift deposit region. This implies that some sort of sudden event occurred with no direct, or measurable, gradation period. Indeed, the drift itself has all the appearances of being the product of some *sudden catastrophe.*

Evidence is also suggestive of rapid geographic position changes of the oceans. On mountaintops all across the world fossilized remains of crustaceans and sea shells have been uncovered. This evidence presupposes that magnetic anomalies do have a direct effect on Climate Change. I have therefore concluded that the sudden onset of Ice Ages is a direct result of some sort of electromagnetic fluctuation. This electromagnetic instability is sufficient to account for many extinction level events and climatic catastrophes one observes in the geological records.

Chapter Twenty

The GCM

Current models based on the "General Circulation Model", or GCM, generate scenarios of climate change rarely longer than a century. The reliance upon GCM generated scenarios may have constrained theories that exclude these longer-term issues of climate change. Thus ideas of 'Abrupt and Smooth' climate change have been predominantly held within a century's boundary.

Despite this narrow window of climatic observation, studies in ice cores have accumulated a significant amount of knowledge which show direct evidence of very large and quite sudden changes in climate. "For example, Ice core records have shown that the mean annual temperature in the North Atlantic region has changed by as much as 10° or 20° over periods as short as a few decades, that is, within one human lifetime (e.g., La Salle, 1993). In some periods, the climate undergoes such changes several times, as if oscillating between two different states before settling into one for a longer period." (Steffen 2001, 4th Annual IPCC.)

Evidently this oscillation cycle has some measure of periodicity even if it is not well understood at the present. And although the current trend is to view these sudden changes in climate as anthropogenic, the majority of these historical climate records preclude any supposed direct human activity. In this setting it is difficult to imagine how any modern society could survive the kinds of changes recorded in say, the Greenland migrations and extinctions. These facts are conclusive and unquestionable.

However, what is not so conclusive are the direct driving mechanisms of so-called Climate Change. If, as estimates reveal, several watts of energy over every square meter were taken away in the Ice ages, what kind of mechanism could have caused this on a global scale? The typical Azazel is quite often remarked to be a change brought about in the Earth's atmosphere due to exacerbated volcanic activity. Yet scientists have been unable to show how minimal amounts of potential cooling could have excited the actual onset of such a large-scale and protracted worldwide event. Most predictive models only show several tenths of a degree in temperature change sustained over only a few years of time.

Other explanations for Ice Ages deal with orbital eccentricity and variations of solar induction at the equator and the poles, however, these atmospheric gas concentrations that are used to substantiate this are effects–*not causes*. A causal basis for Ice Ages needs to be able to explain a marked decrease in these gas concentrations which would include an orbital eccentricity and variation of solar induction as it relates to a proper view of earth core dynamics. There have been many exceptional historical climate periods which also showed significant increases in greenhouse gas concentrations and this during a time of no apparent human activities.

For an explanation about the sudden onset of Ice Ages to be viable it should be able to sufficiently account for both periods of oscillations traced out by the IPCC.

Chapter Twenty One

Ozone Holes

One of the more controversial studies of Climate Change concerns the development of so-called Ozone holes. The methodology ascribed to investigative techniques, similar to the other greenhouse gases, deals primarily with human activity. Here we will discuss an alternative approach to understanding why ozone depletion is so significant in the Polar Regions.

Atmospheric ozone plays an important role in limiting potentially life-threatening ultraviolet radiation by absorbing its harmful rays. In the lower stratosphere it also absorbs upwelling infrared radiation being emitted by the Earth's surface. It is in fact partly due to this interaction that the temperature distribution is smoother in the stable layers of the stratosphere. One finds then an intimate connection between Ozone and lower atmosphere temperatures. (Crutzen, 2004)

Now, while this temperature stability is significant for life to develop as it does it also plays a role in radicalizing chemical stability and other layers of the upper atmosphere. Currently it is understood that UV radiation with a wavelength of about 335 nanometers is capable of splitting an ozone molecule into an oxygen molecule and an excited oxygen atom. This oxygen has enough energy to react with the atmosphere water vapor to produce hydroxyl radicals, OH. (Levi, 1971).

This chemical reaction is written as;[16]

R19 $O_3 + h\nu \rightarrow O + O_2$ (<335 nm)

R2 $O\dagger + H_2O \rightarrow ZOH$

Different gases in the atmosphere are then able to absorb other radiated wavelengths such as oxygen and nitrogen which can absorb up to 249 nm. Through the process of what is called, 'oxygen photolysis', solar ultraviolet radiation of wavelengths less than 240 nm generate stratospheric Ozone. (Photolysis of O2 produces 2 oxygen atoms which combine with an oxygen molecule to form Ozone, Chapman, 1930.)

Originally such a process was believed to be uninfluenced by human activity. By the same token the reverse process of ozone deformation was also largely believed to be independent of human activity. That is until Crutzen in 1970 hypothesized the chemical destruction of Ozone through NO and NO2. The following year in 1971 Johnson and Crutzen independently proposed Nitric Oxide emissions could destroy Ozone. A few years later Molina and Rowland (1974) also hypothesized that Cl and ClO released into the atmosphere from the photochemical decay of chlorofluorocarbon gases (CFC13 and CF2) could possibly deplete Ozone by a similar chain of catalytic reactions.

Based upon these four lionized special interest studies the

---

[16] (h is Planck's Constant and v is a photon frequency)

basic framework for further investigative techniques was laid down. Since the banning of CFCs the human consumption rate has dwindled to almost nothing, yet one still observes a dramatic increase in the deformation process of Ozone. Couple this together with the fact that the Ozone holes are prominently discovered in the least likely of places and we have all the makings of a scientific hoopla.

Up until 1985 it was believed that Ozone destruction would primarily occur over the equatorial latitude range of 30-45 kilometers, however, it was observed at the same time by researchers at the British Antarctic Survey, (Farman, et al., 1985) that the most dramatic ozone decreases were occurring during September to October, primarily in the lower layers of the stratosphere over Antarctica. This finding was, according to Crutzen, "totally unexpected". (2004)

While most scientists who study atmospheric conditions utilize complicated chemical reactions to describe such deformations in polar areas, I believe they are overlooking one important aspect of such depletion (despite the banning of industrial Cl and Br containing gases), which is a combination feedback loop, not of extraneous gas forcing into the atmosphere, but of the waning magnetic field. Now according to the Encyclopedia Britannica, "The existence of the magnetosphere has in all likelihood played a fundamental role in determining the nature of the Earth's atmosphere and therefore in the development of life". (Encyclopedia Britannica, 'The Earth', 2007)

Such a waning magnetic field would have dramatic consequences and could be a causal basis for such a depletion of Ozone, "Exactly in the part of the stratosphere the furthest away from the industrial world, and exactly in that altitude region at which until about 1980 maximum concentrations of ozone had always been found, mainly during the month of September…" (Ibid).

13,000 years ago the Ice Age ended abruptly…with devastating and highly erratic climatic events. This may ultimately be a warning that such a transition is once again occurring. All the signs are here, and we may in fact be witnessing an electromagnetically defined cyclical transition between two overall cooling and heating scenarios. If so, as we approach the apogee, the balance of the earth that we so heavily rely upon may destabilize, initiating a series of climatic domino effects. The critical point appears to be the waning field and the eventual collapse of the magnetosphere.

Conclusion

The burden of this book thus far has been to convince my readers of the possibility that the earth core is not solid. At the same time I've asked you to consider that planets and stars are not gravitationally bound, hence ideas concerning density, mass, and planetary coupling need to be reexamined under consideration of the simple parameters setup in S.E.C.T, quantum core geometry and planetary distance parameters defined by harmonic frequencies. By considering the possibility that the earth core density is charge density, and that prophylactic electromagnetic torsional rotations are causations determining magnetospheric and ionosphere flux, we are able to provide sufficient evidence opening up discussions for further scientific inquiry, and at the same time provide answers to the perplexing anomalies created by adopting a presupposed gravitational constraint. At the same time we are able to allow scientific inquiry to take its natural course unfettered by lobbyists and special interest groups with deep rooted political ties and an elitist bent.

www.ingramcontent.com/pod-product-compliance
Lightning Source LLC
Chambersburg PA
CBHW030450220526
45464CB00006B/2474